The Winged Men of Orcon
A Complete Novelette

by

David R. Sparks

The Winged Men of Orcon
A Complete Novelette
by David R. Sparks

ISBN: 978-93-59329-98-7

Published by

DOUBLE 9 BOOKS
2/13-B, Ansari Road
Daryaganj, New Delhi – 110002
info@double9books.com
www.double9books.com
Tel. 011-40042856

ABOUT THE AUTHOR

David R. Sparks is a renowned author who is known for his fascinating storytelling and insightful narratives. Sparks has an exciting literary style that captivates readers and takes them on thought-provoking journeys through his works, owing to his love for exploring different genres. His writing frequently digs into deep difficulties, giving readers a chance to connect with significant ideas while still enjoying a well-crafted story. Sparks continually exhibits an in-depth understanding of human nature and a flair for developing unique characters, whether he is making suspense narratives, going into the depths of science fiction, or exploring the nuances of human relationships. His imaginative works have had a long-lasting impact on the world of literature, gaining him respect and acclaim from both readers and literary peers. David R. Sparks, a prolific author, keeps continuing to thrill and pleasure readers through his thought-provoking narratives as well as his capacity to take readers to new and creative realms while providing fresh viewpoints on the human experience.

CONTENTS

CHAPTER I
THE WRECKED SPACE-SHIP

When I came to, it was dark; so dark that the night seemed all but fluid with black pigment. Breathing was difficult, but in spite of that, however, I felt exhilarated mentally. Also I felt strong, stronger than I ever had in my life before. I tried to raise my hands, and found that I was handcuffed.

I lay sprawled out on a sharply canted floor of metal, and from outside the house, or whatever it was I was in, I could hear the screeching and howling of the wind. I touched my face with my fettered hands, and the act gave me a feeling of comfort, for the scar on my cheek was still there and I knew that I was myself.

Twisting around, I sat up, and with great difficulty drew a lighter from my trousers pocket. The flame glimmered up. I knew then that I was lying in the control room of a great flying machine!

All about me I saw crumpled human forms clad in glistening gray flying jumpers. It was very, very hot. I thought I caught the sound of waves crashing on a shore. Through a broken port blustered a hot wind laden with an odd odor suggestive of garlic and kelp. It was just as dark outside as in. I stirred about a bit, and found that I was in good shape except for the handcuffs.

A low moan came from behind a bulkhead door at one end of the control room. I listened, and again the sound was repeated. With the lighter still flickering in my hands, I got to my feet. The bulkhead door was jammed, but I found a heavy telargeium spanner-wrench on the floor, and with a strength which frightened me—a strength which could have come only by some upset condition of gravitation—I soon crashed the door open. I had no sooner done it, however, than I forgot about the moan which had fetched me.

What I saw first, hanging on a hook on one wall, was a bunch of keys, one of which readily opened the lock of my handcuffs. Then there was a long-barrelled, gleaming atomic gun, undamaged, and a couple of the new cold-ray flashlights. Free, I caught up one of the flashlights, and placed back on their hook the keys which had opened the cuffs. Then I stooped over each corpse, and confirmed my first impression that two of the dead men were strangers to me, but that I half recognized one.

The vaguely familiar man was clad, under his gray jumper, in the uniform of a rear admiral of the U. S. W. Upper Zone Patrol Division. He wore a medal of high honor, the Calypsus medal. I knew that he was Wellington Forbes, the man who had defeated the planet Calypsus three years before.

Wellington Forbes! And I with him!

I think I may be excused my temporary forgetfulness of the moan which had brought me to Forbes' death chamber. Uppermost in my mind was the manner in which I had been brought here. For it was he, approaching me through the medium of letters and messengers, who had begged, implored me to help him against Orcon, the eccentric planet of my own discovery, the planet which belonged to a solar system at the other end of the Universe from ours. Because of my knowledge of Orcon, with its bubbling seas, its brooding nightmares, and lastly, its queer conduct toward Earth, he had wanted to take me away from my telescopes to fight. And I had refused.

Now I understood how I came to be here.

I knew that this dead man had kidnapped me after drugging me with one of the new amnesiacs. Yorildiside, I reckoned it. And just because I knew that Admiral Forbes had seized me by force, I knew almost to a certainty that I was shipwrecked on that very Orcon which I had discovered two years before.

Iwas enraged at this high-handed treatment. For if danger was indeed threatening Earth from Orcon, my place of all places was at my telescopes. I could do with them, for the civilizations about me, what no one else could. Too, I was actuated by selfish motives. I loved my telescopes and my isodermic super-spectroscopes. And there was still much work I had to do! Already I had discovered three new elements, and that had showed me I was but at the beginning of a knowledge of cosmological chemistry. Forbes! He had brought me by force out here on this beastly little planet whose orbit was like that of a snake with the Saint Vitus' dance! He had taken me to this wretched planet which lay at such a remote end of the Universe that not even explorers had been tempted to visit it!

"Oh, damn the whole business!" I groaned aloud. I was thoroughly angry and bitter.

In a little while I experienced a sudden change of mood. I'd no sooner spoken than a moan came from directly behind me, and I remembered why I'd got going in the beginning, and was ashamed. I entered a small compartment which opened from Forbes' cabin, and discovered immediately three more people.

The strides I had taken made me realize that I had to be careful, for I was indeed endowed with a terrific strength—an extraordinary strength and lightness. One of these three new people was obviously dead, for his neck was broken. The other two still breathed. The first of the two was a short man, a Japanese by the look of him. His arm was broken. The other person was, to my surprise, a woman. She, like the dead Forbes, wore the insignia of the U. S. W. Upper Zone Patrol. Her insignia was that of a navigating officer.

So it was she who had caused the crash!

It was also she who had moaned. My feelings as I lifted her to a bunk were mixed. Being a reactionary, I still felt that woman's place was not in the Army or Navy. Yet I confess that the woman—or girl, rather—was ornamental. She was of the Iberian type. She was beautiful, and looked helpless. Some atavistic trait of the protective instinct in man made me take a little more pains in caring for her than I might have taken with a man.

"Doctor Weeks," were the first surprised words she murmured when I had bandaged a cut in her head and she came to.

Weeks being my name—Frederick Weeks—I grunted and wondered just how much *she'd* had to do with my being here. I noted that the eyes were gray with violet lights.

"You were handcuffed and drugged," she announced wonderingly.

"I was," I answered, "but I'm not any more. Thanks to my own efforts."

She dropped that subject.

"Take me to Admiral Forbes, Doctor Weeks. I am Captain Virginia Crane."

I acknowledged curtly her introduction of herself and told her the admiral was dead. Her cheeks, already pale, grew white. I asked her the number of the space flyer's crew. She said ten. So far, four were dead, three alive, including myself, and the rest unaccounted for, I told her. She winced. In a moment, though, she pulled herself together with a grit which I could not deny, despite my disapproval of her being here.

"I suppose you wonder why you're here," she said suddenly, "and where we are."

"I don't need to be told where I am," I said coldly, "but a little information as to who was responsible for my coming to Orcon wouldn't be amiss. I suppose it was Forbes."

She cut me off with a look.

"It wasn't the admiral." Her really beautiful eyes narrowed. "It was I who planned your abduction and got him to execute it."

"You!"

I drew back. My manner was formal and cold.

And after that I guess I pretty well boiled over. But did it gain me anything? Before I had said half enough to soothe my lacerated feelings, the girl simply shrugged and looked bored.

"Don't be a fool," she ordered curtly. "We needed you, and I, for one, was not going to see your egotistical ideas about an unimportant piece of work—your cosmological chemistry—jeopardize the safety of the world. Oh, I know the government wanted you in your laboratory. But with Ludwig Leider loose on Orcon, and you the only one in our Zone who knew much of anything about the planet, what could you expect?"

Ihardly know what might have happened between us if she had not mentioned Leider's name when she did. The insults with which she had begun had hardly been atoned for by her half understanding of my refusal to join Forbes, and I was still in a rage. Yet, as it was, at the mention of Leider I snapped to attention.

"Ludwig Leider! Here?"

"Yes," she replied significantly.

"But that makes a difference! Why wasn't I told? Why this silly kidnapping?"

She moved a little on the couch and looked at me.

"There was not time to tell you and to chance putting up with further silly arguments on your part. When the secret service detail which had been handling the Leider case brought in word of his whereabouts, there was time only to get a ship specially outfitted for such a tremendous journey and start. We *had* to kidnap you."

I hardly heard her last words. Ludwig Leider—scientist extraordinary, renegade, terrorist. Everyone of our latter day century knew that he was the greatest example of the megalomaniac—the power-seeking genius—which the human race had produced for decades. Everyone knew that he—furious because he had been denied the high position he craved as ruler for life of the united peoples of Earth—had been the leader of the interplanetary struggle which had resulted in Forbes' brilliantly successful attack on Calypsus. And everyone knew that he had escaped from Calypsus. And that, while he was free, there could be no real safety anywhere, either for Earth, which he hated, or any of its allied planets. Leider, here! No wonder I had been observing queer goings on in Orcon!

Somehow I forgot to be angry with poor dead Forbes. Almost I forgot to disapprove of the woman.

"See here!" I broke out. "If your secret service detail was right, and Leider *is* on Orcon, we've got to stop talking and get going. Tell me more about your expedition."

"Do you know," she said presently, "I rather thought you would make quite a leader—and fighter—if you could ever be aroused. As for the expedition, we have only this one ship. It's that kind of a job."

"Oh, suicide party, eh?"

I ignored her remark about my ability as a fighter. I had never aspired to any sort of naval or military leadership.

"Yes," she answered; "suicide party. And I suppose, with our ship wrecked, our admiral dead, and contact with Leider not even made as yet, it's become doubly so. But we've got to do something."

She leaned forward on the couch.

"Our primary objective," she went on, "was to reach Orcon and scout, and then radio information back to Earth. But we also have two tons of the new explosive, kotomite, aboard and are to do damage if we can. What are you going to do, Doctor? The command is yours now."

I was well enough versed in the upper space tactics of our modern navy to appreciate the wisdom which had been used in sending the one ship alone on the expedition, and I could well understand the reasonable hope of success which had been promised. I confess I was staggered to know what could be done, however, now that the admiral was dead and the ship wrecked. As for my having inherited the command, I was even more disconcerted.

"*I* don't know what we're going to do," I said in answer to Captain Crane's question. "I doubt if Forbes would know, if he were alive, and I'm by no means the commander he was. But, as you say, we have to do something. So, since it's a little early in the game to explode the kotomite and call it a day's work, we better declare a truce between ourselves, and then check up on the ship. Come on, if you're able."

She was able.

In the next twenty minutes we found that it was the forward end of the great flier which was damaged, and that while she was in fair shape amidships and astern, she would never fly again. We discovered that the three unaccounted-for men of the crew were lying forward, and found that two were dead and one lived—a radio man named LeConte. He had two ribs broken. Half a dozen atomic guns remained to us, and we found intact one dynamo capable of generating the new cold light in considerable quantities. It was not an encouraging check-up, though. Out of a crew of ten, only the four of us were alive; Captain Crane, the Jap, LeConte, and myself. And all of us were more or less battered. The ship was still habitable, but smashed beyond hope of repair. Around us stretched Orcon—in the control of Ludwig Leider.

I got LeConte, the radio man with the broken ribs, into the small cabin where the Jap still lay and made him comfortable. Then I set the Jap's broken arm. I gave both him and LeConte an injection of penopalatrin in order that their shattered bones might be decently knitted in two or three hours. The Jap presently came to, and I found that he was a civilian like myself, but one who had long been employed on the U. S. W. research staff as a ray and explosive expert. I realized at once that he was the inventor of the kotomite with which the ship was loaded.

All of them, including Captain Crane, told me the story of the crash. Captain Crane hadn't been responsible, after all. Their magnogravitos system had failed in some mysterious manner as they approached Orcon. In spite of the checking effect of their helium pontoons, which had expanded properly when they had come into Orcon's atmosphere, they had slammed into a sea of light and crashed. That was all anyone knew. But everyone suspected that Leider had been somehow responsible.

"I do not enjoy the prospect," Koto said after a glance at his temporarily helpless left arm. "If Leider is able to wreck a space ship before she ever reaches his planet, he has more power than he ever had during the Calypsus war."

Isaid nothing, but simply looked at LeConte, and nodded approval when he muttered something about getting his sending set in shape, if that were possible. We were sitting in the small cabin and Captain Crane was searching my face with those discomforting, violet-lighted gray eyes. I knew she was asking me once more what I was going to do, and I knew that, except that we might fire the kotomite, I could tell her nothing.

We sat on in silence. Then, however, before I spoke about the kotomite, a change came.

All at once I felt the space flier tremble under me. It rocked gently over on one side and began to move. Slowly, but definitely.

Koto and I were on our feet in a flash. Captain Crane stiffened and faced me, waiting.

"What is it?" Koto gasped.

"We'll find out what it is," I flung back. "Miss Crane—Captain—on deck with you. Here, Koto, a hand with one of the guns. We'll take it up out of the hatchway and through the main cabin."

LeConte, I knew, was the one we must be careful of, with his cracked ribs.

"Get to your apparatus," I ordered him, "and stay with it until you get through to Earth."

With that I jumped into the main cabin, stepped over Forbes' lifeless body, and caught hold of the nearest of the atomic guns. I was to be a leader, after all.

CHAPTER II
THE CABLE OF MENACE

It was dark when we gained the deck; as dark as it had been when I first regained consciousness. Captain Crane was attending to that problem, however. As Koto and I floundered with the gun on the slippery telargeium plates of the outer hull, I heard her moving about. Then she uttered a cry of relief, and there came a faint click. Instantly the darkness all about—the clinging noisome darkness of Orcon at night—was shattered.

The blessed rays of our one good lighting dynamo were loosed!

I saw the girl standing braced beside a stanchion, staring over the ship's side.

"Come on, Koto!" I snapped.

I am no fighting man by trade. Nevertheless, there was a kind of instinct which told me to get the gun set up at any point of vantage along the ship's side. And Koto understood.

"There," he breathed after but a few seconds, and from the experienced way in which he touched the disintegration-release trigger with his one good hand, I knew we were ready.

The flier was still moving, slowly and smoothly. She seemed to be half lifted, half drawn by some colossal force. I leaned far out over the rail.

A long, slender, but apparently indestructible cable had been affixed to our stern by means of a metal plate at its end which I guessed to be magnetic. I saw that the cable vanished under lashing waves which broke on a not distant shore, and that we were being drawn irresistibly toward the waves.

The light from the deck brought out dazzling scintillations from a beach composed of gigantic crystal pebbles as large as ostrich eggs. On the beach and grouped thickly all about our hull, swarmed a legion of creatures which—

Well, they were the brood of Orcon. They were the creatures who had given Ludwig Leider refuge and allied themselves with him in his attempt to make trouble for Earth. And they were half-bird, half-human! Their faces, bodies, arms, and legs were human. *But they had wings!* Translucent, membranous structures, almost gauzy, which stretched out from their shoulders like bat's wings. And their skins, as they surged about in the beams of our light, gleamed a bright orange color, and about their heads waved frilled antennae which were evidently used as extra tactile organs to supplement the human hands. I could see instantly that the Orconites possessed a high degree of intelligence. Of all the queer breeds that interplanetary travel and exploration had produced, this was the queerest.

I swung to Koto, who was crouching beside the gun.

"Get rid of that cable before we go under!" I exclaimed.

I had already guessed that the plate which held the cable to our stern was magnetic. It was easy to see that the cable had been fastened there by the Orconites and that our ship and ourselves might be drawn to destruction. I flung myself over to Koto's side to help him with the gun.

The howling wind which had been at a lull as we reached the deck, broke loose again, and, as a gust hit us, Koto, gun, and I were all but swept overboard. The winged legion overside gave loud cries and braced themselves against the gusts. I saw Virginia Crane clinging desperately to her stanchion beside the light switches.

"More light if you have it!" I screamed to her against the wind.

Then Koto and I got the gun going.

My first feeling was one of intense relief. As the thing went off under our hands, and I knew from a faint trembling and a low hiss that the weapon was functioning perfectly, I felt thankful indeed for the instinct which had made me get the gun on deck. It could be only a matter of seconds now until a whole section of the metallic cable was disintegrated completely and until our ship was free.

Breathlessly I watched the greenish atomic stream play along the bright length of the cable of death, and, as Koto and I steadied the gun together, I knew he shared my relief. Despite the howling of the wind, the yells of the Orconites, the continued slow movement of the ship, and the hideous churning of the waves astern, I laughed to myself.

"Doctor Weeks!"

I saw that Captain Crane had gone aft to watch the effects of our fire.

"All right," I bellowed. "What—"

"Nothing is happening back here! Your gun! What's the matter with it?"

I was too startled to answer otherwise than I did.

"Nothing's the matter with it. What's the matter with *you*?"

But the next instant I knew she was right.

"My God, Doctor!" Koto cried, and I knew he had leaped to the same conclusion I had.

Suddenly I brushed Koto's hands away from the gun, and myself directed it so that its ray cut straight across one whole group of the queer creatures on the beach. Then I cursed.

Instead of being cut down, broken like so many blades of grass, not one of the creatures showed that the ray had touched them at all. They only uttered tremendous hoarse sounds that might have been laughter.

I stood up.

"Koto, Leider's found means of protecting both raw materials and living beings against the atomic gun!"

Captain Crane was beside us now, and I saw that she did not need to be told of the disaster. As Koto turned away from the gun, I thought of LeConte below. When the waves closed in on us, he would be caught like a rat.

The shriek of the wind and the crash of waves grew louder. I felt upon my face the sting of spray from the aqueous solution of which the lashing sea at our stern was composed. The cable held, and the ship continued to move. We were barely a hundred yards away from the shore.

All at once, though, a string of both chemical and physical formulae—the last thing a man would expect to think of in such a position—flashed into my mind.

"Here, wait a minute," I thought. "If Leider's done this thing, it means—it must mean—that he's juggled his atomic structures through production in terrific quantities of the quondarium light which I theorized about last year! But he can't have done that without playing hell with the action of magnetic forces from beginning to end! I believe if we take the gun aft and direct it at—"

That was as far as I got with forming words. I flung myself toward the gun and began to drag it to a position aft, where we might direct its ray full force, at close range, against the magnetic metal plate which held the cable to our stern.

"Help me!" I yelled at the others.

Koto was the first to close in. Struggling, slipping, hampered rather than helped by our great strength, we clawed our way aft. A combined lurch of ship and blast of wind threw Captain Crane down, but she staggered up.

We dropped the gun with a thump at a spot where the bulging curve of the stern swelled directly under the muzzle. I grabbed at the trigger just as a new surge of movement brought the flier perilously close to a great, inrushing wall of water which was not water. Koto's face was drawn, and Virginia Crane was staring in horrified fascination at the gun.

Again came the faint trembling of the beautifully constructed mechanism. The green ray leaped out across the blinding whiteness of our light rays. I jammed the muzzle down until the whole force of the atomic stream was spouting against the magnetic plate which held the cable to our stern.

"Look, Doctor! Look!" Captain Crane cried.

But I was already looking.

For an instant a flash of blue light played about our ship. There was a single sharp, crackling sound; and, ringing in the night, an echoing, high-pitched twang.

Koto let out a shout. I took my hands away from the gun.

Backward the twanging cable snapped, demolishing with one touch a score of the clustering Orconites. Into the waves it snapped, and our ship, ceasing to move, came to rest upon the glittering pebbles of the beach.

I heaved a deep sigh.

"What came to me a moment ago," I said breathlessly to the others, "was the idea that when atomic structures are so juggled that they are no longer affected by the gun, all the forces of magnetism, which usually are immune to the atomic stream, are rendered liable to disruption by it. We could not destroy Leider's cable, but we could play the deuce with its magnetic grip on us."

Koto was looking at me wide-eyed, and I saw that his interest was as keen as my own. Even Virginia Crane, scientist though she was not, was interested.

We were in no position, however, to sit still and think. The waves astern and the howling wind were subsiding noticeably, but the inhabitants of Orcon all about us were still creating a great hubbub. Our next obvious move, regardless of what they might do, was to get hold of one of them and make him talk.

After a gesture to Koto and Captain Crane to stay where they were, I ran to a spot on the deck where I had seen a permanent ladder fixed to the side of the ship. Three jumps took me down to the beach, and three more took me into the very midst of the mob.

The confusion brought about by the destruction of the score or so of Orconites by the flying cable, and by our unexpected salvation, all worked for me. And another thing worked for me, too.

These people had great intelligence, but they seemed like sheep when it came to a question of physical, hand to hand encounter. Of rough and tumble fighting with fists they knew nothing—as indeed not many people do in this century, even on Earth. The result of it all was that they shrank back when I charged into them, and not a blow was struck, even when I caught up the nearest figure in my path, swung it over my shoulder, and tore back to the ladder. In two shakes I was standing on the deck again, my prisoner all safe.

"What a creature!" Virginia Crane cried as I presented her and Koto with my struggling but helpless prize.

That was just what I had thought after my first glimpse of the whole brood of them. Close inspection showed, as I had supposed, that the Orconite was a man, and yet not a man. The body, the limbs, the enormous head, the features of the orange-colored face were human; and the chap began to spout excited sounds which were certainly the words of intelligent speech. But also he was winged, and from the orange forehead waved those curious, frilled feelers!

He was clad in a single loose garment of woven cloth which permitted free action for both limbs and wings. A small, flat black box with a mouthpiece into which he could speak, was strapped to his chest in such a position that it was almost concealed by the folds of his blouse. We were to find out presently the purpose of this instrument, but I did not examine it carefully then. As the creature glared balefully at us from his intelligent dark eyes, I glanced over the side of the ship to see whether trouble was to be expected from his fellows.

And for the moment they surged about so much, and made so much noise, that I thought trouble might come. The shouting, however, was caused by their dismay at all that had happened to them, and I saw that instead of making ready to attack they were preparing a retreat. We had whipped them temporarily.

We had thrown them into such disorder, indeed, that in another moment a whole force of them gave proof of their ability to fly, by taking off from the beach. Up and out they swept, out into the intense blackness which overhung the sea behind us. In another moment the whole crew had vanished, and I was glad enough of it.

"Come on below," I said to my two companions. "There's no telling how long Leider will keep his hands off us, and we've got to find out from our prisoner whatever we can."

With that I turned to the companionway, lugging the winged man, and the others followed.

CHAPTER III
IN THE GRIP OF LUDWIG LEIDER

Once we were below, LeConte joined us from the radio room. After taking a swift look at our prisoner, and listening to our account of what had happened above, he reported that the radio had been put out of commission by the crash but could be repaired. All of us then held a hasty conference and decided that since no one was badly in need of rest, LeConte would return to his sending set, Koto would keep a deck watch, and Captain Crane and I would see what we could learn from the prisoner.

From the start it had been certain that the Orconite's strength was not to be compared to our earthly powers. Therefore I made no attempt to bind him, but simply shoved him into a seat in the main cabin of the flier—the room in which Forbes' body still lay—and began to try to make him talk.

I knew that Leider must have some way of communicating with his allies, and I was determined that if he could, I could. But it was uphill work. The creature closed his mouth, assumed a sullen look, and sat tight. He knew what I was after—that I could tell by the expression of his face—but he met with stolid silence all of my attempts to address him in such languages as I knew of Earth and our allied planets. I got nowhere, until, in a manner as sudden as it was unexpected, something happened which ended the deadlock.

The way it happened was this. As LeConte, working in the radio room close off the main saloon, completed a connection which had been broken, he called to us that he was making progress, and a moment later we heard the click of his sending key and the shrill squeal of a powerful electric arc breaking across the transmission points of his set. I realized at once that this did not mean that the set was wholly in order, for the pitch of the squealing arc was too high and too sharp, but I did know that there was hope of establishing communication with Earth soon. And, too, I realized another thing.

The moment that shrill, squealing sound impinged upon the Orconite's ears, he jumped and uttered a cry of pain. There was something about his nervous organism that could not stand these sounds!

"LeConte," I shouted, "close your key again!"

After that the battle was won. By the time I had explained to LeConte why I had given him the order, and he had filled the cabin two or three times with the screech, the Orconite was ready to speak. He trembled in his seat. His mouth twisted with pain, and a look of agony seared his eyes. He burst into fluent Orconese speech. Then he made a swift pass with one hand at the black box on his chest, touched a switch there, and began to rattle his Orconese into the mouthpiece.

The result—well, one might have known that Leider would have found some ingenious means of making the difficult speech of Orcon easy. Out of the small instrument into which our prisoner spoke his hard, rattling words, came a flood of pure German.

An instrument for translating spoken Orconese into spoken German. That was what the little box was.

"Shut the accursed transmission set off!" came from the box in a clear German which I understood readily. "I will talk. Ask what you want to know. I cannot stand this!"

His face still contorted, the Orconese touched a second switch on the box, and indicated that I was to speak at the instrument. I did so, in German. The result was an instant translation into the prisoner's own tongue.

The rest was easy.

"What is your name?" was my first question.

"Hargrib."

"What were you and your people trying to do to us with the cable you hitched to our stern?" I asked next.

"Destroy you."

The whole story was this: In a power house on an island only a few hundred yards off the beach was kept a magnetic cable which Leider had been using in connection with some deep sea dredging apparatus he kept there. When our ship crashed, the order had come from headquarters that the cable be fastened to us and the ship drawn into the sea. I concluded that we had missed an unpleasant fate by a narrow margin.

Quickly Hargrib confirmed our belief that it was Leider who had wrecked our ship while it was still approaching Orcon through space. A ray which had crippled the magnogravitos had been used. So great was Leider's power that, after disabling us, he had even been able to direct our course so that we had crashed on the beach close to the headquarters he had set up for himself deep in the wilderness, away from the cities of Orcon.

The Orconite's free mention of Leider's name and his open admission that the man was king and god in Orcon, made direct inquiry about him easy. Also it was plain that Hargrib, now he had been cornered, would hold nothing back because he believed we would never live long enough to make trouble, regardless of what information we gained.

To state the rest of it briefly, we learned that Leider had come to Orcon immediately after his defeat at Calypsus. He had found ready allies here, on the crazy, distant planet which had been too remote to tempt explorers from Earth until necessity had forced our voyage. The people of Orcon knew science and machinery, and were advanced in every respect. From communication which they had had with other peoples in their own zone of the Universe, they had even heard of Earth and its allied planets. They had lent themselves readily to Leider's fierce plans to make trouble for Earth.

As to what Leider's plan of war was, Hargrib could not tell us much, for his duties kept him absorbed in other work, not connected with the campaign. He stated definitely, however, that Leider had almost completed the development of apparatus which would enable him to strike his blow without ever leaving Orcon. The whole work was being carried forward in tremendous subterranean laboratories and power rooms which had been established in a series of natural caverns only a few miles distant from the desolate beach on which we were lying at that moment. Hargrib said that with the coming of daylight, we would be able to see the mountains in which the caverns were concealed, just as we would be able to sight the nearby island whence had been shot the cable which had so nearly done for us.

At this point my natural curiosity as a scientist made me desire greatly to ask a thousand questions about the planet itself, with its bubbling chemical seas and its erratic orbit, and I did ask a few things. The answers I received confirmed the theory I had already formed that Orcon did not revolve regularly, but had days and nights which might last anywhere from a few hours to a month. I was told—what I had already guessed—that the bubbling fluid which composed the seas changed the orbit of the planet as the nature of the fluid's chemical elements changed.

Also I was told flatly and calmly, as though there were nothing at all remarkable about the fact, that Leider had penetrated so deeply into the chemical secrets of Orcon that he was able to control the coming of day and night. Finally I was told that the planet had a hot, moist climate instead of the frigid one to be expected when any sun was so remote, because of the continued warmth-giving chemical action of its seas.

Icould have gone on seeking information for hours. Captain Crane, however, showed impatience at even the few questions I did ask, and I knew that she was justified. It was my duty to think about the position we were in and the task we had in hand.

I asked Hargrib sharply what was to be expected from Leider now that his cable party against us had failed. And he told me.

The sum of it was that Leider was working eagerly to complete his preparations for the attack on Earth. Although it was he who had sent word from headquarters that we were to be destroyed, he had not paused to attend to the matter himself. Hargrib thought, however, that the failure of the cable party might change this attitude, and expressed the belief that Leider would interview us now before he put us out of the way. He swore, and I believed, that he did not know when or how Leider would come to us or have us brought to him. Also he did not know when or how we would finally be exterminated.

I now asked a series of indirect questions which led me to believe that neither Hargrib nor his master knew of the thing I had been conscious of from the start—that we had aboard the ship an amount of high explosive sufficient to do ghastly damage not only to this section of the coast but to the whole planet of Orcon. I gathered, however, that Leider suspected we were armed against him in some way, and would watch us carefully.

By now daylight had begun to peer in through the ports, a greenish daylight which grew out of the north, and with its coming I resolved on a plan of action.

"I am done with Hargrib," I said suddenly to Captain Crane. "We'll lock him up in one of the staterooms, and after that we'll see if we can't get busy with something that will at least help Earth, even if it doesn't help us."

Hargrib, still terrified by those radio sounds he could not stand, made no protest when I ordered him into the stateroom which had belonged to the ship's second officer, and we were rid of him in a moment.

I now called LeConte from the radio room and Koto in from the deck, and after Captain Crane and I had told them what we had learned, I made my proposal.

The plan was simply that LeConte should continue to work on his sending apparatus until he reached Earth, while Koto, Captain Crane and I set out on a reconnaissance. I said that I hoped to be able to locate Leider's headquarters and learn what method of attack he intended to use against Earth; and that I hoped further that at least one of us would be able to bring word back to LeConte, who could send it to Earth. Finally I indicated that we would see what could be done with our two tons of kotomite as soon as we had made the attempt to send information home. I told LeConte, who would stay with the ship, to fire the explosive himself if anything happened to make him believe that we had been killed while scouting.

I did not fail to point out that since our atomic guns were useless against the Orconites and Leider, we should have to go unarmed on our expedition, and I did not fail to state that the whole effort seemed futile. But the opportunity offered by Leider's present withdrawal was one we could not afford to miss. We were drowning people, I said, and we must clutch at straws. And my friends were good enough to agree.

As soon as the conference was ended, therefore, we disposed of our six dead by the simple process of disintegrating them with one of the atomic guns, and then LeConte returned to the radio, and Koto, Captain Crane and I went on deck to have our first look at Orcon by daylight.

The first thing we saw was the small, rocky islet just off the shore whence had come the cable. It seemed a harmless place now, with only one squat building of stone and no Orconites about, but we were glad enough to turn away from it and look toward the dark and ragged range of mountains which loomed up some five miles inland—the mountains of Leider's headquarters. Not that the sight inspired us with greater confidence. It didn't. But it was good to look at the mountains, because the fact that we were going there meant that at least we should be acting instead of idling.

No Orconite was visible anywhere.

With the coming of daylight—the greenish daylight of Orcon—the sea behind us had calmed until its surface was disturbed only by gigantic lazy bubbles which broke with muffled, thudding explosions. The air smelled of chlorine, iodine, and sulphurated hydrogen, but was breathable. I saw that the principal characteristic of life on Orcon was an organic ability to thrive under almost any climatic conditions. Many of the huge, crystal clear boulders which covered the beach and the coastal plain which led to the hills, were covered with leafless flowers which had immense, leathery petals and sharp, fang-like spines. Other evidences of swift growing life showed on every hand. Ugly, jelly-like creatures oozed about the ship and everywhere else. In places the very rocks seemed ready to come to life.

After one good look about, I issued the order to start. As we clambered down the ship's ladder to the beach and set out resolutely toward the hills, I made myself try to hope, and for a time did muster up a little cheer.

I did not keep it, though. In less than ten minutes something happened which ended our expedition in a terrible manner.

What began it was a long shout which came echoing from LeConte back on the ship. The instant I heard the cry I knew, somehow, that trouble had started. Leider had kept off us as long as we had remained quiet, but at our first move he had gone into action.

While LeConte's cry still echoed in my ears, I swung to face the ship and saw him waving frantically from the deck. At that moment I also had a queer impression that the sunlight was growing brighter on all the glittering rocks, and that some new feeling was creeping into the air.

"Doctor Weeks!" LeConte cried across the distance between us. "Come at once!"

Terror had laid hold of the man. Captain Crane, Koto and I began to run to him.

"What is it?" I shouted.

"I don't know," came the thin answer. "I almost had Earth when my whole set went to pieces. Come, quickly!"

"We will, if we're able," I muttered to myself, and said aloud as I ran in the gigantic bounds possible on Orcon: "Koto, Captain, do you feel anything queer in the air as if—as if—"

I never finished. Suddenly Captain Crane screamed and flung out her arms to me with the gesture of one about to fall.

"Doctor Weeks!" she gasped. "Frederick, help me!"

And that was all. Before she could choke out another word, before I could do more than clutch at her, she had been caught up by an invisible power, caught up straight into the now dazzlingly brilliant green air, and swept away from us as if she were a feather in a tornado.

It was over before realization could sink in. Nor was her departure all. From the ship came a ringing yell, and as LeConte, in the distance, clutched a stanchion as if for dear life, the whole battered, glimmering gray shape of the flier moved, shivered, and in a flash was caught up and whisked away as easily as had been Virginia Crane!

"He's got us!" I sputtered as I turned to Koto. "He was only waiting until we started to march against him."

"God, yes. Horrible!" he muttered.

Then *his* kindly yellow face went white. Even while I stood looking at him, he, too, was swept away into space.

When my turn came, it was as if implacable fingers took hold of my wrists, the front of my coat, my shoes. I distinctly remember thinking that after all the peace we'd had, something as astounding as this was almost bound to have happened. The glittering boulders of the coastal plain fell away, and I felt myself being whirled through space. The speed was taking my breath away. A ringing came into my ears, spots floated before my eyes, a nauseating light-headedness swept me, and I lapsed into unconsciousness.

CHAPTER IV
IN THE CAVERNS OF ORCON

Icame out of it to find myself lying on my back upon the rocky floor of a cavern more lofty than any cathedral. The air was warm and charged with a pungent, almost mephitic odor. Blue light filled the vast subterranean place. I heard the far-away, droning throb of machinery. Crackling sounds like static on a vast scale ripped back and forth at intervals.

Neither Captain Crane, Koto, nor LeConte was in sight, but wherever I looked as I twisted my head slowly, I saw winged Orconites staring at me. They stood back against the walls of the cavern chamber, their wings folded, the antennae on their orange foreheads waving gently. None was close, but all watched with cold, intelligent interest. I decided that I was in Leider's headquarters, a closely guarded prisoner. It was to be supposed that Leider had brought us here, as Hargrib had said he might, to interview us before he finished us off.

Fear for the others laid hold of me, but I was still too dazed and giddy to get up and look for them. I lay still, trying to remember everything.

"He waited until we made an aggressive move," I thought, "and then he did *something* to us. He did something which brought us shooting through the air here to his headquarters!"

After I had progressed so far, it did not take me long to realize what method Leider had employed to fetch us to the caverns. Nor did it take me much longer, once I was sure of the method, to roll over heavily and begin to yank the metal buttons off my coat. Since the many guards—fully twenty of them—made no move to interfere, I did not stop until I had torn every button off my clothing, dumped from my pockets every object which had a scrap of metal on it, and even dug the metal eyelets out of my shoes.

What had happened was that Leider had simply readjusted the forces of his damned power houses so as to yank us to him, ship and all, *without* the medium of a magnetic cable. What he had done was to direct at us a magnetic current so terrific that, taking hold of the few odds and ends of metal on our persons, it had snatched us bodily through space. And the ship, too! It was stupendous; incredible.

Full consciousness had returned by this time, and fear possessed me even more completely than it had before—fear for what might be going to happen to Earth and fear of what might already have happened to my friends. The Leider who had planned the Calypsus war had had no such gigantic powers as these. As thoughts of Virginia Crane and the others increased until they filled my whole mind, I sat up on the floor of the cavern and then rose slowly to my feet.

The guards never relaxed their vigilance, but they made no move as I moved; they only stared, and I ventured to call out.

"Captain Crane! Koto! LeConte!" I shouted loudly.

No answer came. Since the Orconites still did not prevent me, I began to walk swiftly down the length of the great, echoing cathedral cavern, toward an abutment of rock which jutted out from one wall, separating the room I was in from another. Again I shouted, and the whole place rang with echoes, and my fears grew.

But all at once fear vanished. I knew that the worst had not happened and that I was not to be left alone.

"Doctor Weeks!" It was Koto's voice, and it came from behind the abutment of rock toward which I was hurrying.

"Koto!" I yelled and entered the next cavern and saw it all.

He was lying stretched out on the rocky floor of an underground room as vast as the one I had left behind me. He was unhurt, and he was waving to me! Captain Crane, just waking up, was stretched out beside him. Our ship, a colossal bulk of battered, gleaming metal, had come to a lighting point some fifty yards beyond them. LeConte was sitting on the deck, staring groggily at me.

Guards were posted all around the walls of this new cavern, and those I had just walked away from now came crowding in to join their fellows, but none spoke to us or held us back. In another thirty seconds LeConte had slid down from the ship, Captain Crane had stumbled to her feet, Koto had flung an arm about me, and we were all babbling together.

I will not attempt to tell of our feelings during that interval. But the reunion did much for us. When I had returned to consciousness, it had been with the thought that our puny scouting expedition had been wrecked before it had begun, and that all else had been lost to us. Now the mere fact that we were together once more changed my attitude suddenly and completely.

"Defeated?" I asked myself, and as I gripped the warm hands of friends I knew that we were not defeated at all. Rather it seemed that everything we could have hoped to gain was won.

The penopalatrin I had injected in Koto and LeConte had mended the former's broken arm and the latter's cracked ribs, so that none of us was in any way disabled. And we seemed to be free within limits. And our ship was here in Leider's caverns—our ship laden with two tons of the most terrific explosive science had ever created. And the Orconites, though they might be suspicious, knew nothing of our weapon.

Now that hope had sprung to life again, I knew that the opportunities open to us were huge. We were in great trouble, and whatever we did would probably not be easily done, but there was a strong chance that we might yet strike a blow that would help the peoples of Earth in their hour of need.

It was not necessary to explain to the others all that was passing in my mind, for I could tell by their expressions that they were comprehending the possibilities as clearly as I.

"What's Leider up to?" Captain Crane asked after a while.

"He's brought us here to put us through an interview," I answered. "He hasn't sent for us yet because he's busy getting ready for his war. Also, since he's a Prussian all the way through, he's probably ignoring us in the belief that his absence will make us more impressed with his mightiness."

"Yes, but what are we going to do while he ignores us?" she snapped back.

"Quite a lot," I answered, and turned to LeConte. "What are the chances of getting word to Earth?"

"Impossible," he said, shaking his head. "The set was wrecked when the magnetism—or whatever it was—took hold of us."

"All right. Never mind it." I looked at Koto now. "Koto, what do you have to do to fire your explosive?"

I was sure now that the thought had already been in their minds, for Captain Crane and LeConte nodded and Koto smiled.

"The kotomite," he answered, "is packed in telargeium drums in the ship's hold, and protected against being exploded until oxygen is admitted to the drums and force applied. It was our original hope to land on Orcon, deposit the drums, and fire them by a time fuse. The quickest way now would be simply to place one of our atomic guns in the hold, turn it loose, and get out. The stream of the gun would in a very short time disintegrate the drums to admit oxygen, and would at the same time set off the explosive."

"Good," I said shortly, and without more ado glanced about the cavern to look over the situation with regard to the forty or so Orconites whom we had been ignoring, and who had ignored us, ever since we found each other.

They were standing motionless against the walls, eyes alert, ugly antennae waving, but with their arms folded across their chests. There seemed to be no reason why we should not all march boldly to the ship, climb aboard, and forthwith do the work that was to be done there. I had, however, a feeling that our task was not to be so easily accomplished, and was not long in discovering that the feeling was correct.

The moment I told the others to come with me, and we all started to walk toward the ship, the whole encircling force of Orconites began to move silently forward. When we were within a few yards of the ship's ladder, a tall lithely built Orconite who seemed to be captain of the guard, flopped his wings, shot across the cavern, and dropped down before us. Into the instrument on his chest he rapped a word of Orconese which was translated instantly into the German.

"*Verboten!*" was the word.

Forbidden! The Orconites were not taking any chances with us. It was discouraging, but no more than I had expected. It simply meant that if we were to be interfered with, we should have to do something about the interference.

I quickly began to work out a plan.

First of all I shrugged at the captain of the guard and turned back from the ship as though his refusal to let us aboard was of no consequence. Next I spoke to the others.

"Come on," I said in a normal voice. "Don't make a fuss now, but pull back, from the gangway."

They saw, I think, that I was planning something, and we retreated together, with the result that the Orconites ceased to threaten and once more fell back to the walls of the cavern. Their captain flew over and joined them.

"Ithought for a moment," I said, "that we might tell the captain that Hargrib was locked up in the ship, and so furnish an excuse to get aboard. But that isn't good. Some of the Orconites would surely go with us, and in that case it would be next to impossible to get at the kotomite properly. What we need is at least a couple of minutes which will be uninterrupted. We'll leave Hargrib right where he is, and get access to the ship in another way. We'll fight for it!"

"Fight?" Captain Crane shot a glance at me, and I saw that the idea appealed to her.

"So far as I can see," I said quickly, "Leider hasn't armed his guards with any unique weapon, but has merely left them to watch us. And the Orconites don't know how to fight! Think of the ease with which I got away with Hargrib last night. When it comes to dealing destruction with scientific weapons, their power is appalling. When it comes to a slugging match, they are only so many sheep. And Leider's forgotten to take that fact into account!"

I felt really sure that the guards were not armed with some mysterious weapon we could not see, and Koto felt the same.

"Doctor, you're right!" he exclaimed. "Leider's made a mistake! He's forgotten what damage can be done by physical strength, and left us alone with a mere flesh-and-blood guard. There are forty of the Orconites and their leader, and only four of us. But we have strength that they never dreamed of possessing. It makes the odds almost even!"

"Right," I snapped. "And they will be even altogether if we can get hold of some clubs."

Koto and the others looked doubtful at that, but I had been thinking hard of the problem all the while we were talking. I motioned unobtrusively toward the end of the room, where a tunnel, blue-lighted and lined with curious, glittering dials like ammeters, gave entrance, evidently, to another great underground chamber. On the floor of that tunnel, close to the entrance, lay a pile of heavy stalactites of some mineral which resembled jade. The spikes had seemingly been cleared off the tunnel roof and left to be carried away. They were pointed enough to be used for stabbing, and looked heavy enough to make stout clubs.

Captain Crane smothered an exclamation as she glanced at the pile, and Koto and LeConte smiled.

Our conversation all this while had been carried on with seeming casualness, and not even the leader of the Orconites showed suspicion. More than ever I felt that neither they nor Leider would be prepared to defend the ship against a sudden physical attack.

"The weak point for us," I said, "is that we'll have to make an awful row, and the alarm will go out, and eventually some weapon will be brought out to stop us. But if we work quickly, there's a good chance that we can finish everything before Leider is able to step in with some devilish freak instrument. Take it easy until we've got the clubs, and then cut loose for all you're worth. Captain Crane, it's a great pity you're a woman. In all this you'll simply have to—"

I did not finish. Something in the look she gave me stopped me quite, and somehow, whether I would admit it or not, I knew she was as fit as we were. By this time we were strolling away from the ship toward the tunnel.

Blue-lighted, brilliant, the opening loomed larger as we approached. The same sounds of static on a vast scale which filled our cavern, filled the tunnel, but the place was deserted. The pile of jade spikes shimmered right at the entrance. A few of the guards behind us sauntered at our heels without speaking, and the dozen or so about the tunnel closed in toward the opening, but no restraint was put upon us.

"We seem to have the freedom of the place and the key to the city!" was Captain Crane's dry comment.

"Yes," I answered. "I'm pretty sure it's going to be a case of lambs led to the slaughter. Looks as if—Oh, good Lord, look!"

At the moment when I spoke those last words, we had approached to within thirty or forty feet of the pile of stalactites, and from the quick movement which eight or ten Orconites made ahead of us, drawing themselves up in a line across the tunnel mouth, I knew that we had almost reached the limit of our freedom. But it was not that fact, or the movement of our guards, that brought the exclamation from me.

"Look!" I cried again, even though I knew each of the others had seen as clearly as I.

From where we were walking slowly forward, it was possible to see clear down the tunnel to the tall, lighted cavern beyond our own. In the center of that cavern, with her nose pointing toward a wide tunnel down which showed a glimmer of daylight, rested the long, needle-like, bright hull of the most beautifully designed space flier I had ever seen.

We did not need to be told that this was Leider's own cruiser. A ship of such magnitude and exceeding beauty could have been nothing else.

The guards knew we had seen and were aware of our excitement, but contented themselves by standing fast in the line they had already formed across the tunnel. We advanced another few yards.

"Mother of Mercy!" LeConte whispered, almost in awe.

"There's a chance for us!" Koto gasped. "A chance! We'll set one of the guns going in the hold of our own ship, and then—"

Captain Crane's face was flushed with intense excitement, and her fingers were moving as though she felt the delicate controls of the space ship under them even now.

"Could you pilot it?" I asked.

"*Could* I! Give me the chance!" she cried.

"All right," I snapped, "we will!"

And in that second I enlarged my plans to take this gorgeous new development into account.

"Fight to take the cruiser," I ordered. "Captain Crane, Koto, LeConte, get aboard as soon as you can cut your way through. I'll take care of our ship and the kotomite at that time and join you, if possible. Come on!"

Thus was it decided. Thus did we enter our fight with an outlook as utterly different from our original one as hope is different from despair. Our discovery of the cruiser had been almost accidental, a thing which might never have taken place except for our trip to get the spikes of jade. Surely such a happy accident had never happened before!

The moment I gave the command to go ahead, and we started to run, all of the ugly, bird-like faces of the Orconites across the tunnel became convulsed, and the creatures commenced to howl at us. Before we hurled ourselves against the line, swift reinforcements shot through the air over our heads and joined them, and the temporary uncertainty which had held them gave way, so that they met our advance with an advance of their own. But we did not care.

A few smashing blows which I delivered with my fists served to bring screams of agony from the several creatures immediately about me, and as one or two staggered and crashed to the floor, the others gave way a little. In a moment I was through the line to the pile of stalactites. And the others were through with me.

"Here you go, Koto!" I cried, and stooping down in spite of the jostling bodies and clammy hands that tried to prevent us, I caught up one of the long, needle-pointed, heavy stalactites. As I shoved it at him and snatched another for myself, Captain Crane and the others armed themselves.

By this time every Orconite in the heavy guard was on the spot, and the whole mass was all over us, gasping, burbling, flapping their wings, fighting to clutch at us with their hideous orange hands and waving antennae. Decidedly the fight was on, and I was forced to admit the fact that, though these creatures might be sheep, even sheep have power. But the first skirmish was already won, and I had faith that we could win the real battle.

I balanced my peculiar weapon in my hand to get the feel of it, then brushed aside a pair of sucking paws which were trying to take it from me, and plunged the spike clean through the body of the man who held me.

He fell without making a sound. I regained my weapon by planting my boot on his chest and wrenching it free.

I swung the spike like a club and crushed two heads with a single blow at each. A downward blow served almost to hack a long, clutching arm from an Orconite's body. With four men out of the struggle, I looked to see how my companions were faring, and was assured by a single glance that they were as well off as I.

Encouraged greatly, I met an advance of pressing, jostling bodies by a return to my original technique of stabbing. I stabbed every time a hand reached out to hold me, and if I did not take a life with each stab, I at least drew a spout of greenish-colored blood.

It was not a nice business, any of it, especially as the Orconites were as fearless before our onslaught as they were powerless. But it had to be done. We were fighting for far more than our own lives.

The blue-lighted corridor with its rippling sounds of static and its gigantic ammeters became worse than a shambles. We walked upon, stumbled over, wallowed amongst the piled corpses of the slain, whose master, knowing more of the science of destructive warfare than any other being in the Universe, had nevertheless forgotten that it was still possible for mankind to fight with their hands.

Such a fight could have only one ending.

When the end came I saw that Virginia Crane was splashed with the ugly blood of the Orconites from her smooth forehead to the soles of her flying boots, but she was unhurt. The rest of us were likewise blood-stained and uninjured. We were all too excited to feel tired. The moment the pressure about us began to relax, she surged toward the waiting cruiser at the end of the tunnel, and I shouted to Koto and LeConte.

"Go and help her, you two! I'll do the work on our ship!"

They did not question my order, but obeyed.

There were only ten or a dozen of the winged ones left now, and when the two men leaped after the woman, it was easy for me to fight a jabbing, slashing battle which not only protected the retreat, but enabled me to work my way slowly toward our own ship and its kotomite.

With Leider's cruiser already headed toward the tunnel which led out from the underground hangar, I knew that it could be taken into space with a minimum loss of time. I believed that I could get an atomic gun going in our hold quickly, too. My hopes rose high as I darted a glance over my shoulder and saw Captain Crane and Koto taking, three at a time, the gangway steps which led to the deck and control room, with LeConte directly behind them. Now there were only seven guards left instead of a dozen, and those were at last showing signs of being cowed. I cut down two, and gave a great bound which carried me away from the others in the direction of our wrecked ship.

No sooner, though, did I tense myself for a second leap than I felt a nerveless sensation in my knees, as though the bones had turned to butter, and knew that my high hopes had budded too soon. Instead of leaping, I staggered on for two short steps, then stopped because I could stagger no farther. Looking back at the cruiser, I saw that LeConte, still on the gangway, had stopped also. Captain Crane and Koto were making weak, despairing signs at me from the entrance to the control room. Both of them looked as sick as cats. I heard a laugh, a shrill, rasping sort of laugh, from the forward end of the bright cruiser, and I looked in that direction.

I saw a short man, bald headed, with frog eyes peering at us from behind thick prismatic glasses. He was clad in baggy green overalls, and was slowly waving in our direction a glistening metal tube which he held in both hands. From the end of the tube emanated a purplish light.

"You were clever, my good young friends," he chortled, "to think of fighting with your hands, but you were not quite quick enough. Not to-day goes anyone in my cruiser! What do you think of the enervating ray, heh? Ingenious, not? Ludwig Leider discovered it. I am Ludwig Leider. You shall come with me and with your own eyes watch the de-energizing of New York and Paris and Berlin. For I am ready to do away with your paltry Earth now!"

I felt the last energy ooze out of me and I sunk, all in a heap, on the floor of Ludwig Leider's cavern.

CHAPTER V
DEATH IN A BOX

New York. We did see it with our own eyes. The instrument through which we gazed was like a metal box with a ground-glass top and a mesh of slender wires leading away from the table on which the box rested. Leider touched a button amidst a long row of buttons on the table. All we had to do after that was to look at the ground-glass plate, and the picture was there.

We, in Leider's private laboratory on Orcon, saw the crowds of a mass meeting of some sort in Union Square, saw a boy and a girl kissing each other in the shadow of bushes in Central Park, saw a little fox terrier watching with only one eye open.

We could not speak, any of the four of us, as we stared at that very simple box which wrought miracles. I stood still, thinking of the things which had happened after our capture, when the cruiser had already seemed to be in our grasp.

First of all, Leider had restored our energy to us by the simple process of turning off the ray which emanated from the tube in his hands. Then a veritable legion of Orconites had come to the cavern in which the cruiser rested, and we had been marched through the very heart of the power rooms, with their hum and clack and dazzle of mighty machinery, to the laboratory. That was all.

The Orconites had left us outside the heavy doors of the private room, but, just as there had been no opportunity to attack while they marched with us, Leider gave us no opportunity to harm him while we were alone. Though he had forgotten once the damage we could do in a fight, he was not going to be fooled again. He kept the great table of the box between ourselves and him, and his wary hands were always closer to a certain row of control buttons than ours were to his.

It was he who broke at last the silence which had fallen as we watched New York from Orcon, and his voice was loud in the hushed laboratory from which the noises of his subterranean power houses were shut out.

"Sit down," he commanded, "and keep away from the table and the reflector."

Then, when we had taken chairs beside the table, he began to speak to us.

"That little dog you saw—I have it in my power to withdraw from him in one second all the energy which makes him run, jump about, live. That I can do by touching controls here at my table without even leaving this marvelous, marvelous room." A frown crossed his forehead above his pop-eyes, and he exclaimed with swift anger, in a croaking voice, "And what I do to the little dog, I can do as easily to the whole population of your loathsome Earth!"

I looked up at him where he stood with the table between us, and at length found my tongue.

"And of course you will do it, you swine!" I burst out.

His momentary anger had passed as swiftly as it had come, and, ignoring my epithet, he rocked smugly on the balls and heels of his feet and smiled.

"Ah, Herr Doktor," he answered contentedly, "I will destroy Earth, of course! For who has better cause than I, whom Earth would not accept as her master? All of the people there will lose the power to move, and they will die. I am ready now, in the uttermost degree. After you so neatly but uselessly saved yourselves from drowning last night, I finished. As easily can I de-energize the peoples of Earth as I can you—the four of you—if you should make the move to harm me."

Captain Crane was staring first at Leider, then at me, and her cheeks were gray and ghastly looking. Koto and LeConte were both sitting tight in chairs beside our own, watching me rather than Leider. I looked over the shelves, the whole complex apparatus of that incredible room, but saw no weapon of any kind. And my hands were useless because *his* were so close to the damnable controls.

"But what becomes of Earth itself, after our peoples are gone?" I asked presently.

Leider shrugged and his eyes twinkled behind the thick glasses.

"Herr Doktor, you are a brilliant man. Amongst the most brilliant, I should say, of any who on the Earth have labored. Yet of science you know less than a child. What should I do with Earth except to sit here in my own room, and, with the anarcostic ray, reduce its solid structure into stardust which will drift away into space like the smoke from one tiny match? Pouf! like that."

I looked at the table, at Leider's wary hands. I knew that the man was ready, even as he had said, to do away with Earth. I guessed that we would die, too, when Earth was gone—probably here in this room. And it seemed likely that the destruction would begin at a not distant moment, for there was some quality of fanatical evil lurking even now in Leider's face.

Then, however, I stiffened in my chair very suddenly indeed. If I could find a way to get close to the box on the table without rousing Leider's suspicion, the outlook might not be so black!

"Leider," I exclaimed all at once, and there was a vigor in my words, "it's all very well for you to be saying these mighty things, but do you know what? I don't believe you can draw the energy out of the human race or disintegrate the Earth, either!"

Ithink if I had kicked him I could not have surprised him more. Which was exactly what I had hoped to do.

"You—you do not *believe*?" he said, incredulously.

"No I don't!"

"Ach, Gott!" A black fury overcame him. Hideous fury. He was already standing beside the table. Quaking from head to foot, he pointed savagely at the box. "Get up and look into the reflector!" He choked and his voice rose to a scream. "Get up! Stoop close to the reflector and watch! Watch there, I say!"

The thing which had launched me on my course of action was the fact that the picture-making box was not screwed to the table. The only thing which held it there was the soft mesh of wires!

With a concealed gesture to the others to stay still, I rose, placed my hands on the table close to the box, and leaned forward as though to look at the glass.

"It shall come now!" Leider yelled, and at that moment took his eyes off me, while he reached with a rage-palsied hand for the twinkling line of buttons.

The instant he looked away from me, I gave a tug which jerked the heavy box away from its wires as easily as a weed is plucked from soft earth. As I made the move Leider looked up and screamed. His hand, already reaching for the buttons, darted forward. But the instant had been all I needed. Before the darting hand ever reached the table, I struck Leider a sharp blow, and hurled the box to the floor.

In a moment more the others were around me. The box was shattered to matchwood. Leider was lying on the floor behind his table with one arm doubled limply under him and dark blood welling from a forehead gash which I hoped went as deep as his brain.

Koto and LeConte kicked open the laboratory door and shot through. Captain Crane and I jumped after them.

CHAPTER VI
THROUGH THE DARKNESS OF ORCON

Gongs clanged, blue lights flashed on and off with the lurid glare of sulphur pits burning in hell, and screaming, winged Orconites, all mixed up together, pelted toward us as thickly as the snowflakes of a blizzard. I don't suppose the destruction of one little mesh of wires had ever created such a disturbance before.

Leider's cruiser rested in the hangar two caverns away.

"Play hide-and-seek with them!" I shouted against the turmoil.

The initial wave of the attack struck us as we tore from the laboratory corridor into the first power room. Captain Crane went down under the onslaught of what must have been a hundred Orconites, and it took all the tearing strength of Koto's, LeConte's, and my hands combined to burrow through the piles of creatures who covered her, and get her out. By the time she was on her feet again, a new legion was at us.

I had not, however, suggested hide-and-seek meaninglessly.

While the others fought, and wildest confusion reigned, I pulled off my coat, flung it aside, and crammed myself into a loose, one-piece costume of Orcon which I tore off a corpse. Then I fought while my three companions repeated the operation. We succeeded in confusing the mob to such an extent that we were able to work our way through the fringes of the melee and move clear across the first room, before we were recognized.

The alarm of our escape, though, spread into the next room almost as soon as we reached it, and a foolish attempt we made to keep bunched together and get through with a dash, betrayed us before we got well started.

Now it was a case of being drowned again by a sheer deluge of men. While the Orconites pawed me, tripped me, and otherwise discommoded me, I broke necks, dug out eyes, tore quivering antennae from foreheads until I felt as if I had been doing nothing else for hours. And those beside me were doing the same. Yet always more bladder faces rose in front of us, and more wings beat down from above. Not even our supreme strength was great enough to stand it.

Out across the bleeding, crumpled bodies and the teeming swarms beyond, I saw as through a red mist the glittering, whirling maze of Leider's wondrous generators, and began to curse to myself.

For the steady pressure was forcing us slowly back toward the machines and toward the rugged, high wall of the cavern beyond, and I knew that once we reached the wall we could retreat no farther and must stand there to fight until we were completely exhausted. I drew closer to Virginia Crane and did what I could to help her with her main group of assailants while still battling my own.

Oddly enough, I was remembering how, when she had been caught up by the magnetic current that had brought us here, she had cried out to me, calling me by my given name.... The recollection filled me with a queer emotion, partly rebellion and partly—something else. In the crisis we were facing now, I somehow lacked my wonted power to shun femininity.

Side by side we struggled against our enemies, tearing at them with our whole strength, yet always we were driven closer to the wall which would finally stop us.

"Oh," she finally gasped, "I—didn't want—to die!"

"No," I answered through set teeth as I hurled down an Orconite only to be confronted by two more; "but I'm afraid—we must. Well, we've done away with Leider, anyway."

"Yes," she choked. "That's—something."

Koto and LeConte were as hard pressed as we. Then, as we fell steadily back into a passage between two of the vast generators, back toward the solid wall of the cavern, a queer thing happened.

Despite the fact that LeConte was embroiled with a dozen winged men, his face became crinkled with a broad grin!

"Watch!" he yelled suddenly, and I *did* watch.

We were within a few feet of the driving gear of one of the generators. Quick as a bolt of lightning, LeConte caught a deadly firm hold on one of the ugly, squawking orange-skinned creatures, raised him into the air, and there held him poised while he swung around to face the generator.

Genius!

There was a shriek, then a thousand shrieks. Impelled by the Frenchman's tremendous heave, the winged man shot forward and struck full, with a splashing sound, against the terrifically revolving armature. A thunderbolt seemed to explode in our faces. All in that room, we as well as the Orconites, reeled dazedly back. A stench of seared flesh and short circuited wires smote our nostrils. Darkness—smothering, thick, absolute darkness—settled over us.

"Come on!" LeConte shouted amidst the blessed inkiness of it, and I felt him tug at my hand. Captain Crane's hand slipped into my other, Koto caught hold of her, and we started forward.

Genius indeed, this stroke of LeConte's.

Clinging stoutly to each other, we pushed through the meager, floundering opposition which was all that was offered in the intense darkness, and began to forge swiftly ahead. Ten yards ... a hundred. A slight decrease of the sounds of crying and panting and of confused flopping wings told us we had passed through the arch which separated the wrecked power room from the hangar.

"Captain," I whispered as we battered against some confused and helpless Orconites and flung them aside, "could you make anything of the control system on the cruiser before Leider got us?"

Virginia Crane said vigorously that she had.

"The light switches are all on a board to the right of the entrance door. The other controls are as readily accessible."

"Leaves us in something of a position!" I whispered.

The hand which she had placed in my own tightened its grip. I heard LeConte grunt with satisfaction as he pressed forward. I began to figure on ways and means of getting to our wrecked ship alone after the others were aboard the cruiser.

We crossed another fifty or sixty yards of the darkness, and found fewer of the badly shaken Orconites in our path. Now, in that thick obscurity, I sensed that we were nearing the magnificent, tapering hull with its fish-scale sides.

"Come on!" I urged unnecessarily. I kicked into several of the yielding bodies left from our first fight, before Leider had taken us, and in a little while the feel of cool, smooth metal under my hand told me we had reached the gangway.

"Up you go, Captain!" I snapped, and as she clutched the slender rail of the gangway and plunged upwards, "LeConte, you next. Koto—"

But Koto laid a firm hand on my arm.

"No, I do not go."

We stopped where we were. The noises of pursuit were still around us, and I could have slugged him for making a delay.

"You fool, get aboard!" I roared.

But it did no good.

"No."

"Get the motors started!" I called to Captain Crane. "LeConte, you help her." Then I turned to Koto and in the dark waved a fist under his nose. "You idiot—"

"No, my friend," he laughed at me. "You killed Leider. LeConte put out the lights. Captain Crane will pilot the ship. Now it's my turn. You will pardon my insubordination, but you will also please to hurry up the gangway before I knock you unconscious and throw you up. Damn it, it's my explosive, anyway, isn't it? Who has the best right to fire it?"

With that he whirled away from me.

"Don't wait!" he called over his shoulder.

I laughed at him and sang out the order to Captain Crane to stand by. As for myself, I remained standing on the small platform at the foot of the gangway.

The moment Captain Crane flipped a switch which flooded the control room and a score of ports along the hull with golden light, I thought the yells which rose from the other room and the far side of this one would blow the roof off. By the time we felt a quiver run through the hull, and heard the sweet, deep-throated hum of the gigantic power plant, a mob of Orconites had formed for a new attack. It was hideous that we could not wait for Koto in darkness, but the light was essential to Captain Crane's preparations, so there was nothing to be done. I felt that Koto's chances of getting back to us were one in a thousand.

Yet suddenly, as I still clung to the foot of the gangway, LeConte thrust his head from the control room door and yelled at me to hang on tight. At once the ship moved forward, and, rolling easily on her ground gear, swung left and lunged toward the swooping mob of Orconites.

Handling that space flier in the cavern was like trying to navigate a one-hundred-thousand-ton freighter in a pond. But Captain Crane did it—she whom I had once accused, to myself, of misnavigating and wrecking our other ship. The Orconites had formed themselves in a dense group. We went into them, mowed them down, stopped under the great arch which led to the inky black power rooms, backed up, and, as the screaming lines reformed, crunched terrifically into them again.

By this time I saw in the corridor leading to our old ship, where the darkness was only partially broken by our lights, a dark-headed grinning man who was bent nearly double with the speed of his running.

"He's coming!" I howled.

"He's coming!" LeConte echoed to Virginia Crane in the control room.

And again the miracle of the hundred-thousand-tonner in the pond was performed. Again the cruiser backed up and swung around. We headed toward Koto, straight toward him.

There still were droves of Orconites to contend with. Flocks of them had taken to their wings, and were filling the whole upper reaches of the cavern, now that a juggernaut had the floor. They had spied Koto and were swooping toward him. But they could not seize him without coming to the floor, and they could not come to the floor without contending with the juggernaut.

Now the cruiser seemed to swoop. I saw a swirl of wings all about, battering down and down about the Jap; then I clung to the gangway rail with one hand and reached far out with the other toward our friend.

He leaped, and I felt the warm contact of his hands gripping my arm. I gave a heave, and landed him on the steps as neatly as a fisherman ever netted a trout.

"All clear!" I screamed up the gangway.

It was not until we were on the deck, and the cruiser was gliding magnificently forward toward the shaft which led outside to space and light, that Koto spoke. But when he did, his words had significance.

"It's done!" he panted. "The gun is firing against the drums!"

We dove into the control room, and LeConte banged the outer door shut and jammed huge catches, battening it down for our flight through space.

"Get out as fast as you can!" LeConte panted on, speaking now to Captain Crane as she headed us gently into the tunnel. "The kotomite's due to go off the second the first drums are disintegrated."

I dropped limply on to a seat beside the pilot and sat still.

We passed through the tunnel in five or six seconds. In another five seconds, we had not only taken off, but had worked up a formidable speed. We barely felt the explosion when it came. But on the instrument board in front of Virginia Crane, gleamed a little box with a ground-glass top, and in that we saw, as by a magic, what happened on Orcon.

First the mountains which topped the subterranean power houses were lifted off. Then the whole planet rocked. Finally the caverns were inundated by the deluge of the sea which, in the beginning, had so nearly swallowed us.

Orcon was not destroyed, but we knew even then that such of its inhabitants as might remain alive would not soon again dream of making an attack upon Earth.

On the way back, as Earth took form and grew round in the interminable reaches of space ahead of us, I got on well with Captain Crane. It started when she asked me if I were still so cocksure that woman had no place in the U. S. W. Upper Zone Patrol, and I was forced to answer that I was not. After that, one thing led to another.

We were photographed together when we landed beside the colossal, metal-roofed hangars of the Long Island station of the U. S. W. The snapshot was published in that afternoon's tabloids under the caption: Betrothed.